A DROP OF RAIN IN ASIA: A BRIEF INTRODUCTION TO TRADITIONAL BURMESE MEDICINE

DR. U WIN KO

EDITED BY JOHN HAMWEE AND MATTHEW CLARK

MAHABONGO WEENY INTRODUCTIONS
VOL. 5

Booklet design by Matthew Clark (Mahabongo.com)
Edited by John Hamwee and Matthew Clark
Copyright © Dr. U Win Ko (2011)
Published by Lulu.com
First published in 2011
Second edition (2013)

ISBN 978-1-291-50026-4

Contents

	Page
Preface and Acknowledgements	1
Introduction by Kirsten Germann	2
Section 1: Historical background	3
Section 2: Four systems (*naya*s) of medicine	8
2.1: *Desanaya* and the five elements	8
Diagram of the method of eight types of diagnosis	15
2.2: *Netkhattanaya* and *Weizzadharanaya*	18
Section 3: Traditional Myanmar herbal medicines	19
Section 4: Different treatment systems	21
4.1: Massage treatment	21
4.2: Hot fomentation (hot compress)	22
4.3: Inducement of perspiration	24
Section 5: Integration of TMM with TCM and acupuncture	26
Section 6: The Department of Traditional Myanmar Medicine	30
References	33

Preface

This booklet by Dr. U Win Ko provides a brief introduction to the traditional medical systems of Myanmar (formerly Burma), which are called 'Traditional Myanmar Medicine' (TMM). U Kyaw Than Win, also known as Dr. U Win Ko in academic circles, graduated from the Institute of Traditional Medicine in Myanmar in 1980. He worked for seventeen years in the department of traditional medicine as a senior lecturer and as a physician in the Traditional Medicine Hospital, Mandalay. Under a WHO fellowship programme, he studied bone setting at the Institute of Luo Yang Orthopaedics. With thirty years of practical experience, U Win Ko is now head of the acupuncture unit at the Watchet Jivitadana Hospital and is vice-chair of the Myanmar Traditional Medicine Practitioners Association, Mandalay region.

Acknowledgements

It would not have been possible to publish this booklet without the help of Professor Aung Myint, who is Director General of the Department of Traditional Medicine, Myanmar; Dr. Kirsten Germann, who organized the presentation of my paper in England; and Dr. John Hamwee (Pholapyae) and Dr. Matthew Clark, who edited my text and arranged for its printing. The project would not have come to fruition without the help, support and assistance of all my benefactors, to whom I am deeply grateful.

I am also very grateful to Dr. Dudly Kent for introducing me to the British Acupuncture Council, conducting interviews, and for other valuable help; and to the British Acupuncture Council for inviting me to present this paper, which forms this booklet, and for sponsoring my trip to England.

Dr. U Win Ko

Intoduction

Traditional Myanmar Medicine is indeed a drop of rain in Asia, possibly more like the many drops of life-giving monsoon rain. This form of medical treatment is rooted in antiquity and Buddhist tradition, and is akin to Ayurveda, the system of traditional Indian medicine. The methods used are herbal and mineral medicines as well as unique massage techniques, which are very powerful in restoring health. Traditional Myanmar Medicine has a lot to offer to the world.

This may be the first written introduction, or one of very few, explaining to a wider audience the history and principles of this medical tradition, which is largely unknown to the world outside Myanmar.

I first met Dr. U Win Ko and his colleagues in 2006 and we have worked together ever since. Having both witnessed and personally experienced Traditional Myanmar Medicine, I can attest to its power and effectiveness for many conditions. It is like a hidden gem that shines in the dark, yet to be discovered by the rest of the world. I hope that this booklet may be the beginning of this gem coming into the light and becoming recognised and more widely studied and used for the benefit of all people.

The seeds for the auspicious meetings with Dr U Win Ko and his colleagues, and finally for this booklet, were laid while I was working at Wachet Jivithadana Hospital on its yearly acupuncture project. This hospital now has become a true leader in combination medicine, offering western medicine, Traditional Myanmar Medicine and acupuncture, all to a high standard.

For more information on the acupuncture project, please go to
www.myanmaracupunctureproject.org

Kirsten Germann Lic.Ac.MBAcC

Section 1: Historical background

Myanmar, formerly Burma, lies in south-east Asia. Its area of about 676,587 square kilometers is bordered by China, Laos, Thailand, Bangladesh and India. Its population is 60 million, seventy per cent of whom live in rural areas. The main ethnic group was formed from the union of Mongoloid tribes, *kshatriya* (warrior-king) settlers from India, and local, ethnic people.

Traditional Myanmar Medicine (TMM) was established by about 850 BCE and was influenced by the culture of neighbouring countries and in particular by Ayurvedic medicine which was brought from India by the *kshatriya*s. The other main influence was Buddhism, also from India. The Lord Buddha is said to have made numerous visits to Myanmar during his life.

1. Tagaung era: 850 BCE–400 BCE

This is the earliest recorded era of Myanmar history. Tagaung is an ancient city. The story goes that the kings of Burma (Myanmar) were descended from an immigrant Indian tribe which arrived in Tagaung around 850 BCE, several centuries before the time of the Buddha. Their leader, Thado Jabudipa Dhajaraja, established Tagaung as his capital and founded a second dynasty, of seventeen kings. The lord Buddha visited Burma with a host of followers. He made numerous visits to Sandalwood Monastery at Pwintbyu and left behind his sacred footprints.

In this era, although archaeological evidence has not yet been found, descriptions survive of the construction of a golden palace and of ceremonies of ascent to a throne, which included a feast that was first tasted by two cats and two men, most probably as a precaution against potential poisoning of the food. There are also indications that there was at that time some knowledge of toxins, antidotes and herbal medicine.

2. Sriksetra era: 400 BCE–900 CE

Sriksetra became the capital of a number of city-states such as the old city of Vishnu, Hanlin, Maing Maw and others. Excavations of the 'old city' of Vishnu have revealed a variety of agricultural products, including sesame, beans, various kinds of vegetables, sugarcane, and milk products like curds, which were probably used medicinally. Two types of coin have also been found: one type shows the *srivatsa* ('holy family') symbol on one side, and the *svastika* ('moving sun') on the other; the second type shows the *srivatsa* on one side, and the Buddha's throne (*pitha*) on the other side. The second type of coin was used not only for trade but was also worn as a medallion or a pendant to ward off evil and to bring good luck to the wearer. This indicates that *Weizzadharanaya* (see Section 2.2) medical knowledge was widely accepted in society at this time.

In this era, there is also some evidence of tattooing used as a form of medicine performed by Ari monks. These monks were acquainted with Buddhist scriptures and wore dark brown robes and conical hats. Their exact identity is uncertain, but they may have been Mahayana Buddhists.

3. Pagan era: 900–1300

It was in this era that written language was developed in Pagan, and when there is the earliest evidence for TMM, found at Mg Chit Sa's farm land to the east of the Ananda Pagoda, in the form of votive tablets dated 1113–1160 CE. These tablets, which name medicinal plants, are the earliest writing found so far in Myanmar. Another stone inscription, from Saws Hla Wan, dated 1236, gives the title 'Thamar' to someone who knows medicine. The Myanmar script had its source in the Brahmi script from India, which is the earliest of the Indic scripts. The consonant and vowel signs closely approximated the sounds of Myanmar's own languages and the script became well established by the 11th century. The Ari monks of this period were astrologers and soothsayers and also provided medical treatments.

4. Pinya and Sagaing era: 1312–1354

The well-known treatises *Lawkaniti* and *Hitawpadetha* come from this period and were written by Saturingabala who was a high official of the Myanmar royal court. Some medical terms such as *haritala* (yellow orpiment), and *vithala* (bitter melon) are found in these treatises.

5. Ava era: 1364–1733

The stone inscription of Thetnwekyaung has four faces and specifies the cause of disease. The well-known monk poets Shin Mahasilavansa, Shin Maha Rahttasara and Shin Aggasamadhi composed numerous poems which included medical knowledge.

Taungphilar Sayadaw moved to Ava at the invitation of King Anaukphet Luan Min and complied substantial medical treatises such as:

(a) *Konchardat Kyan* ('treatise on the explanation of details of the elements'),
(b) *Sampannadat Kyan* ('treatise on the conjunction of elements'),
(c) *Kammajayote Kyan* ('treatise on the origins of the physical make-up of a being'),
(d) *Mahanarikonchudat Kyan* ('great treatise on the details of the elements'),
(e) *Ahtadatukane Upadetha Kyan* ('treatise on the principles of the eight elements').

6. Konbaung era: 1752–1885

Taungtwinsayadaw Khingyi Phyaw, whose title was 'Shin Nyana', complied numerous medical treatises, for example:

(1) *Dwadarasi* treatise,
(2) *Dwadathasinta Nakkhat* treatise,
(3) *Adicappa* treatise,
(4) *Sanawuti* treatise,
(5) *Athitidat* treatise,
(6) *Angavijjatika* treatise,
(7) *Kawaisara* treatise.

The *Khway Saung* treatment system, which was written by U Myat Tun in this period, has treatments for gynaecology, paediatrics, fever, piles, smallpox, chickenpox, ulceration, diarrhoea and psychological problems. The terms used in this book indicate that they were derived from Ayurvedic medicine.

The venerable monk Maunghtaung Sayadaw, also called Nyanabivansa Damma Thenapati, also complied numerous medical treatises and was urged by the king to send scholars to India to bring back medical knowledge. The king sent a delegation to India in 1785, which brought back six texts, which were then translated by Maunghtaung Sayadaw. He also compiled a glossary of 700 medicinal plants.

A very popular medical treatise called *Naya Nga Twe* was written by a royal clerk, U Kaung, in 1805 and has been in continuous use among practitioners of Traditional Myanmar Medicine ever since.

An Italian citizen, Mr. Sanjamano, arrived in Myanmar in 1783. He lived in Myanmar for twenty-five years and studied its traditions, religions, administration, education, and its health and medicine. He wrote extensively about the common diseases of the time and criticised the fact that medicine was practised without examination or licence. He also stated that the practice of vaccination for smallpox was brought from Arakan in 1798 during the reign of king Bodaw Paya.

In the third period of the Kongbaung era (1857–1885) many treatises of the Ayurvedic system were translated into Myanmar by monks such as Lin Ka Rama, Srilanka Ashin Damaratana, Bangla Sayadaw, and Ashin Ravinda Mahtae. They translated *Bethitsa Myintzutha,* which referred to seventy-five treatises of Ayurvedic medicine. Another famous book, *Utubawzana Singhaha*, was complied by U Pho Hlaing, in which is set out how to live a healthy life in accordance with the seasons. He also complied a detailed account of anatomy called *Kayanupassana*. U Po was a dietary expert and based his system on the five-elements theory (see Section 2.1).

It was in this period that U Hmont invented and described the 'six diagnostic system'. This system was based on the diagnosis of 'hot' and 'cold', which can be the cause of any disease in human beings. He successfully treated King Min Don's bloody bowel movement and rheumatism and the king awarded him the title of 'Tikissaca Bithetcaraja' (specialist clinical practitioner) in 1866. He also compiled the *Taung Tha* system of medicine. It has twelve sets of palm-leaf inscription, and has been used by TMM practitioners ever since.

By this time, TMM (especially the *Taung Tha* system) had become well developed and had spread all over the country. King Min Don encouraged this development and appointed royal physicians in the following categories:
(1) descended from a lineage of practitioners;
(2) having prominent skills in both theory and clinical practice;
(3) having passed a test set by royal physicians.

On the recommendations of a councillor, a treasury councillor, a royal clerk, an administrator and an officer of court, the king would appoint a royal physician with the titles 'Naymyosetcaraja' and 'Mahabisetcaraja', with a salary of fifty-three *kyat*s (silver coins) a month, an outfit, and a special bag.

Lao Si, Lao Sang Cheng and Jen Chao came from China 1869 and presented to the king a book entitled 'Gandalayit Medicine', which explained how to use the bones of various animals, such as from the tiger, elephant, and rhinoceros.

Traditional Myanmar massage was also popular. The king appointed ten masseurs for himself and seven masseuses for his family, each with a salary of thirty-five *kyat*s a month.

In the reign of king Thi Baw (the last king of the Myanmar dynasty), U Hment, a famous physician, died in 1877, having passed on his knowledge of the *Taung Tha* system to his followers: U Hmont, U Pan Tha, U Pho Min, U Kae, U Chan Tha, his son, U R Sara (a monk) and others.

7. The British Empire: 1885–1948

In this period Myanmar was ruled as part of the British Empire. There were numerous practitioners and monks who were expert in TM medicine. Among them: the monk Mahasangharaja Taung Khwin (who complied *Authada Sanghaha*), Saya Hmet (who complied 'Explanations of the *Taung Tha* system'), Thonse U Pan Tha (who complied *Da Tuvinissayarupaganda*), U Chan Tha (who complied the 'Truth Signs of the *Taung Tha* system'), U Pho Minn (who complied *Abidama Dat Kyan*, *Rupavineissara,* and *Paramatta*), U Kae (who complied *Jivita Vattana*), Galoon Saya San (who complied *Laukhanuzu*) Khanbu Monk (who complied the 'Clinical Practice of Khanbu'), Myo Pyin Kyi Sayadaw A Shin Sandavansa (who complied *Vinineissalra*), Manlesayadaw (who complied *Alanbuta*), Leti Sayadaw (who complied *Raugandaradipani*) and the monk U Eindriya (who complied *Thabavadama Daturasa*). These are the most famous practitioners and compilers of medical treatises.

The British government was also interested in TMM and in 1930 organized a committee of inquiry into indigenous systems of medicine. The committee inquired for two years and presented a report, which recommended the setting of examinations, the training and the registration of practitioners, and the setting up of a clinic, hospital and a herb garden. The report also recommended that among the subjects to be taught in a training school should be: (1) *Desanaya* (the *Taung Tha* system), (2) *Beithizzanaya* (the Ayurvedic system), (3) astrology, and (4) occult science (*Weizzadharanaya*). (See Section 2 for an explanation of these systems.)

8. After Independence: 1948–1962

Myanmar (Burma) became independent on the 4th January of 1948. On the 23rd July 1952 the government set up a committee to modernise Traditional Myanmar Medicine. It has the following terms of reference:
1. to implement the report of the committee of enquiry into TMM;
2. to register practitioners of TMM;
3. to organize the Myanmar Medical Council;
4. to draft the law for the Myanmar Medical Council.

This resulted in the statutory organization of Traditional Myanmar Practitioners. In accordance with this law of 1953, the government opened nine traditional medicine clinics in Yangon and Mandalay. The registration of practitioners was undertaken between 1955 and 1962 and resulted in 22583 practitioners being legally registered.

9. The period of revolutionary government: 1962–1988

The revolutionary government took power on 2nd March 1962 and withdrew the law of 1953. On 11th March 1962, it reorganized Traditional Medicine Practitioners. It set up examinations to test the qualifications of practitioners in December 1962 and built the Institute of Traditional Medicine in Mandalay. This Institute was opened on 31st January 1976 and the twenty-five-bed teaching hospital was opened in 1977.

10. The period of the State Peace and Development Council: 1988–2010

The State Peace and Development Council has upgraded TMM and established it as the official national complementary health system. It has opened 237 township clinics, fourteen hospitals, one university of Traditional Myanmar Medicine, one national herbal garden and many herbal gardens across the country.

Section 2: Four systems (*naya*s) of medicine

Sources of Traditional Myanmar Medicine
Traditional Myanmar Medicine has four main sources. Two of them, *Beithizzanaya* (Ayurvedic medicine) and *Weizzadharanaya* (medical alchemy and spiritual powers) have been in use since about 850 BCE. *Netkhattanaya* (medical astrology) is also very old. By contrast, the fourth source, *Desanaya* was invented in 18th century and is the system that is still widely used at present time.

Section 2.1: *Desanaya* and the five elements
Desanaya is based on Buddhist teaching. It holds that our body is an aggregation of mind and matter. An individual body is made up of a physical aggregate and a spiritual aggregate. The physical aggregate is known as *rupetkhanda*. The spiritual aggregate is divided into four groups: *vedanakkhandha* (sensation), *vinnanakkhandha* (consciousness), *sankharakkhandha* (volitional activity) and *sannakkhandha* (perception).

According to *Desanaya* the body is composed of four main elements: *tejo* (fire), *pathavi* (earth), *vayaw* (wind), and *apo* (water). *Akasa* is a kind of fifth element, which can be thought of as the space between tissues and cells and the openness of pores when perspiring.

(1) *Tejo* (fire)
This element is concerned with heat and cold. When it is heat it is called *ushna tejo* When it is cold it is called *sita tejo*. It governs the digestion of food and the maturation of cells. It combines with the other elements in order to animate the body and generate change. Clinical manifestations of *ushna tejo* are:
1. thirst,
2. excessive perspiration,
3. insomnia,
4. discomfort when lying down,
5. cold feet but warm trunk,
6. tender skin,
7. withered physical appearance.

Sita tejo will manifest in the opposites of these.

There are four kinds of *tejo*:
1. *pripachaca tejo* which digests food and warms the body. It is seen as residing between the stomach and the colon;
2. *jirana tejo* which governs the maturation of cells. If it is in excess the cells decay too rapidly and the person ages quickly;

3. *santapapana tejo* which can cause fever with aching, a burning feeling and alternate types of fever;
4. *daha tejo* which can also cause fever and a high temperature.

(2) *Vayaw* (wind)
This element is concerned with movement. All physical action and all the movements of the organs are governed by *vayaw*. There will be no obvious signs of the *vayaw* element in a healthy person.

Clinical manifestations of *vayaw* are:
1. frequent urination,
2. palpitation,
3. changes in bowel movement,
4. changes in menstrual circle,
5. miscarriage,
6. difficult labour,
7. cough,
8. heaviness of limbs,
9. mood swings.

There are six kinds of *vayaw*. These are:
1. *asasapasasa vayaw* which governs the rising and falling of the breath,
2. *udingama vayaw* which governs rising,
3. *adogama vayaw* which governs descending,
4. *kusisaya vayaw* which governs the cavities of the body (especially the abdominal cavity),
5. *kauhtasaya vayaw* which governs the gastrointestinal region,
6. *angaminganusari vayaw* which flows around the body, especially the limbs.

(3) *Apo* (water)
This element can be seen in all the varieties of body fluids, such as blood and tears, and governs permeation, cohesion and blockage.

Clinical manifestations of *apo* are:
1. swellings and enlargement,
2. weak blood,
3. weak urination,
4. scanty menstruation,
5. dry skin.

(4) *Pathavi* (earth)
This element governs objects and gives them their characteristic hardness or

softness, strength and stability. There are traditionally twenty manifestations, such as in bones, teeth, skin, heart and so on.

Clinical manifestations of *pathavi* are:
1. slowness of defecation or urination,
2. inadequate sweating,
3. rough skin,
4. stiff joints,
5. aching body.

(5) *Akasa* (space)
This is the element that governs space in the body; for example, the space between cells and in the nostrils.

In general, in men, the right-hand side of the body is under the influence of *pathavi* and *apo*, while in women these elements govern the left-hand side of the body. The opposite side, in each, is under the influence of *tejo* and *vayaw*.

The causes of disease
The Buddha said: '*Sitehnapi rupapti uhnarpi rupapti*', which means that all physical phenomena are changed by heat and cold. Hence, all of diseases are caused by some disorder of *tejo* (normally called a 'rebellious state' of *tejo*). Imbalances in, and between, the other elements are seen as consequent or subsidiary causes of disease. This leads to the question: when does the *tejo* element become rebellious?

One instance is when *tejo* in the body is contained and blocked in by *pathavi* and *apo* at the outer layer of the body, and especially the skin. It can happen because *akasa* is weak. By analogy, if the windows of the kitchen are closed, then the heat of the cooker is not dispersed and the temperature inside the room becomes intolerable. Similarly, if the pores of the skin are obstructed, then *tejo* can become rebellious and that can then cause disorder in the other elements.

Diagnosis of the cause of disease depends on an examination of the four systems of discarding: sweating, defecation, menstruation and breaking wind. If there is no sweating, it may be because there is some obstruction of *apo* and *pathevi* in the skin. If there is constipation there may be an excess of internal *pathavi*. If there is no regular menstruation or no breaking of wind, there may be a weakness of *vayaw*. All these cases will lead to an excess of *tejo*. In acute conditions, there will be aching: headache, upper-backache and so on. Many chronic organ conditions are caused by a long-term underlying imbalance in *tejo*, influenced by lifestyle.

There is saying in Traditional Myanmar Medicine: 'All human beings are conditioned by *kamma* (volitional action), *citta* (consciousness, mind), *utu* (climate) and *ahara* (nutrition)'. Accordingly, these are the fundamental causes of disease, and are known as the remote causes. They lead to imbalance in the five elements, and such imbalance is known as the intimate cause of disease. For example, a patient's lifestyle may lead to an imbalance in *pathavi* (earth), which represents the internal organs, and which manifests in loose stools and tiredness.

This classification allows the practitioner not only to reach an accurate diagnosis but also to give the patient advice to maintain health.

Kamma (volitional activities)
All activities can cause disease. For instance, overwork can lead to weakness of *apo* and *pathavi*, whereas a lack of exercise can cause obesity, which is seen as an excess of *apo* and *pathavi*. Then there are the physiological consequences of those actions, which can be seen as the result of *kamma*. If there is an excess of *ushna tejo* (heat) there will be scanty, dark urination. If there is excess of *sita tejo* (cold), there will be profuse, pale urination.

Citta (mind, consciousness)
The Buddha said that 'all actions are led by mind'. So, for example, if there is depression, there will be a physical expression of that mental state, perhaps in premature ageing.

Mental states
In Traditional Myanmar Medicine there are fifty-two types of mental states. The following are most often seen in clinical practice:

1. *Cetana*: is the force which makes actions happen. If *cetana* is strong, then there is a seriousness and intensity of action. It can create either *ushna tejo* (heat) or *sita tejo* (cold).
2. *Lobha*: means attachment to the enjoyment of sensual pleasure. There is an old Myanmar saying: 'The more one gets, the more one needs'. The desire to do good—for example, to give to the poor—is not *lobha*, which can create *ushna tejo* (heat).
3. *Dosa*: means hatred or anger. Both emotions can create *ushna tejo* (heat).
4. *Issa*: means jealousy or envy. It also can create *ushna tejo* (heat).
5. *Macchariya*: means stinginess, meanness, or being unwilling to let others have the same prosperity or the same dignity. It can create *ushna tejo* (heat).
6. *Alobha*: is the absence of greed. It can create *sita tejo* (cold).
7. *Adosa*: means the absence of hatred. It is loving-kindness and forgiveness. It can create *sita tejo* (cold).
8. *Sati*: means mindfulness or awareness, particularly of the arising and passing away of mind and matter. It can create *sita tejo* (cold).

Utu (climate)
There are five types of climate in Myanmar. These are: wind, cold, summer-heat, dampness and dryness. In Traditional Myanmar Medicine, our bodies have a particular systemic relationship to the climate. For instance, in the hot

season the inside of the body will be considered cold. As the natural *sita tejo* (cold) cannot exist outside it will take refuge within the body. Similarly, in the cold season, the inside of the body will be hot, which accounts for the fact that digestion is easier in the winter. And if a disease becomes worse during winter, it indicates an excess of *ushna tejo* (heat).

Ahara (nutrition)
Poor diet is an important cause of disease, as are artificial chemicals in food. However, Traditional Myanmar Medicine sees a good diet as one that is balanced among the elements and tastes, as follows:
1. the sweet taste corresponds to *pathavi* and *apo* - and is cold,
2. the sour taste corresponds to *tejo* and *pathavi* - and is cold,
3. the salty taste corresponds *tejo* and *apo* - is also cold,
4. the spicy taste corresponds to *tejo* and *vayaw* - and is hot,
5. the acrid/astringent taste corresponds to *pathavi* and *vayaw* - can be hot or cold,
6. the bitter taste corresponds to *vayaw* element - can also be hot or cold.

As sweet food will increase the *pathavi* (earth) and *apo* (water) elements in the body, so it will also decrease the *tejo* (fire) and *vayaw* (wind) elements. Spicy food, which is hot, will increase the *tejo* and *vayaw* elements and disperse *pathavi* and *apo*. It also amplifies *akasa* (space) as well.

Compatibility and opposition
Pathavi and *apo* are seen as the same in key ways; they are closely associated with *sita tejo*—the cold state of *tejo*—and are classed as compatible elements. Similarly, the *vayaw* and *akasa* elements are seen as compatible and are associated with *ushna tejo* (the hot state of *tejo*). This basic discrimination is used in diagnosis, for if the clinician finds an excess of *sita tejo*, there will also be an excess of *pathavi* and *apo*. Equally, if there is an excess of *ushna tejo* there will be an excess of *vayaw* and *akasa* as well.

The theory also contains the following classification of opposites:
pathavi opposes *vayaw*,
apo opposes *akasa*,
sita tejo opposes *ushna tejo*.

Hence, if there is excess of *pathavi* and *apo*, there will be deficiency of the opposite elements, *vayaw* and *akasa*. And if there is excess of *ushna* (hot) *tejo*, there will be weakness of *sita tejo*, *pathavi* and *apo*.

Heat and cold
Temperature and heat are different. Every healthy child and adult has a normal body temperature of 98.4° F. Although their body temperatures are the same, different people may have a different heat or calorific value. An adult male, for

example, has a higher calorific value than a child because of their respective body weights. Therefore *tejo* will be different in each of them, irrespective of temperature.

Other forms of energy (for example, light or magnetic fields) can be seen as *tejo*. The acid-base equilibrium of modern medical science can be seen as a balance between *ushna* and *sita tejo*. An excess of acidity in physiological function is a kind of *ushna tejo*, and an excess of alkali is a kind of *sita tejo*.

The human body responds to environmental changes through the activity of *tejo*, for example by sweating to release heat. Digestion is better in cold weather because *ushna tejo* retreats from the skin, goes deep inside, and produces more calories. The modern science of metabolism draws a distinction between catabolism and anabolism, which corresponds to the two aspects of *tejo* (and more specifically between *pripachaca tejo*, one of the four kinds of *tejo*, which is responsible for digestion and warming of the body; and *jirana tejo*, another of the four kinds of *tejo*, which is responsible for maturation of cells and the process of ageing). A good balance between the opposing aspects of *tejo* gives health and a long life.

The discrimination of cause and effect

In Traditional Myanmar Medicine there is a hierarchy of causes and effects of disease, and diagnosis involves them all. At the top level are the four basic causes of disease already mentioned: *kamma* (volitional action), *citta* (consciousness, mind), *utu* (climate) and *ahara* (nutrition). Beneath the level of these four primary causes, there are other levels of potential causes of disease: imbalance in *tejo*; and then, beneath that, imbalance in one of the other elements. There is further discrimination that may be applied according to whether the imbalance is internal or external. For example, organs such as the heart, kidneys and stomach are identified with *pathavi*; but externally *pathavi* is seen as *apo* in the skin, the outer layer of the body. Further discrimination may take account of excess and deficiency of a particular element, on the one hand, and its tendency to dispersion or coherence on the other.

Diagnosis

Pathavi and *vayaw*, its opposing element, are seen as the internal elements. As already explained, if there is an excess of *pathavi*, there will be deficiency of *vayaw*, and vice versa. Internal *pathavi* corresponds to the twenty internal organs, and although there are numerous signs of organ functions, those of the stomach, brain and colon are considered vital.

The signs of a normally functioning *pathavi* are:
(1) good appetite and digestion (stomach function),
(2) good sleep (brain and heart function),
(3) good bowel movement (colon function).

By contrast, if the element is imbalanced there will be:

(1) too much eating or loss of appetite,
(2) too much sleeping or insomnia,
(3) constipation or loose bowels.

Apo and *akasa* (its opposing element) are seen as the external elements. If there is an excess of *apo* there will normally be deficiency of *akasa* because it becomes contained and obstructed by the excessive cohesiveness of *apo*. Similarly, if there is excess of *akasa*, there may well be weakness of *apo*.

The signs of an imbalanced *apo* are:
(1) headache, muzziness,
(2) no perspiration,
(3) halitosis.

By contrast, if the element is imbalanced there will be the opposites of these.

Negative symptoms can be seen as the result of these elements being out of balance. For example, constipation and difficult bowel movements indicate an excess of internal *pathavi*. Oedema is due to *apo* in excess and, by contrast, profuse sweating is due to excess of *akasa*. Infrequent, dark urination shows that *tejo* is in excess, whereas frequent urination is due to an excess of *vayaw*.

The same is true of outward behavioural signs. Patients who are active and energetic are seen as displaying *ushna tejo*; however, if there are outward signs of slowness and passivity, then that indicates the predominance of *sita tejo*. If a male patient has muscular tension on his left side then that is a sign of *ushna tejo*; while if those symptoms are evident on the right side, then that is a sign of *sita tejo*. It is the other way round for women. If symptoms worsen in colder weather, then that is a sign of *ushna tejo*; if they worsen in hotter weather, then that indicates *sita tejo*. Similarly, if hot food aggravates the symptoms then that is a sign of *ushna tejo*, but if it is cold food that makes them worse then that indicates *sita tejo*.

This diagnostic method, derived from the *Taung Tha* system, is called the method of eight types and categorises disease into eight groups. The treatment principles based on this categorisation will either tonify weakness or disperse excess.

Method of eight types of diagnosis by a natural principle of particular dietary tastes and the requisite medicine

name of diagnosis	symbolic name	causal element	predominant elements of pathological state	internal *pathavi* excess	internal *pathavi* weak or collapse	external *apo* excess	external *apo* disperse	taste of medicine to be prescribed	cardinal signs according to basic principles
1 *ket-kha-la* hardness	H1	ushna (hot) tejo	ushna pathavi abandhana	excess pathavi		cohesive apo by excess		cold, bitter salty, sour	headache, no sweat constipation, body ache, stiff joints and muscles, fever with high temperature
2 1st *vihtambhita* swelling 1	H2	ushna tejo	ushna pathavi akasa	excess pathavi			dispersion of apo	cold, bitter salty astringent	constipation, burning sweating, scanty urination
3 2nd *vihtambhita* swelling 2	H3	ushna tejo	ushna vayaw abandhana		collapse pathavi	cohesive apo by excess		cold, sweet oily, sour	loose stool, headache muscle-burn, cramp, scanty urination, facial oedema
4 *prissava* melting by heat	H4	ushna tejo	ushna vayaw akasa		collapse pathavi		dispersion of apo	cold, sweet oily, astringent	loose stool, sweating, weakness, anemia, palpitation, scanty urination, insomnia
5 *sangahita* coherence	C1	sita (cold) tejo	sita (cold) pathavi abandhana	excess pathavi		cohesive apo by excess		hot, spicy bitter	constipation, profuse urination, stiff joints and muscles, heavy limbs, excess sleep
6 1st *byuhana* enlargement 1	C2	sita tejo	sita pathavi akasa	excess pathavi			dispersion of apo	hot, bitter, astringent	constipation, good appetite, sweating, good sleep, profuse urination
7 2nd *byuhana* enlargement 2	C3	sita tejo	sita vayaw abandhana		collapse pathavi	cohesive apo by excess		hot, oily, spicy	oedema on legs, loose stool, pale urine, heaviness
8 *peggarana*	C4	sita tejo	sita vayaw akasa		collapse pathavi		dispersion of apo	hot, oily, astringent	cold loose stool, watery diarrhoea, weakness, dizziness, sweating

Explanation

Upper-four diseases (from H1 to H4) are heat-effected diseases due to *ushna tejo*, whereas lower-four diseases (from C1 to C4) are cold-effected diseases, due to *sita tejo* respectively.

Traditional Medicine Practitioners (TMPs) should adjust dietary tastes according to the medicines they prescribe to a patient. If the taste of drugs are not appropriate for the excess elements and its required taste, TMPs should recommend the necessary herbal plants, such as decoctions, extracted juice, or some fruit juice, or hot or cold water.

For example, if one drug has primarily a cool, salty and sour taste, disease (H1) should be treated with a decoction of bitter gourd, which has a cool and bitter taste. On the other hand, neem, which has a hot, bitter taste, can be given for disease (C1) along with a decoction of ginger. These preparations can result in a hot, bitter or spicy taste, which can supplement the required medicinal taste in the body.

Examination of the pulse
The pulse provides detailed information about the state of the five elements. The pulse on the radial artery is the one that is mainly used but other pulses can be taken in other circumstances.

In ancient times, the physician who was expert in pulse diagnosis could tell not only the present problems of a patient but also his or her previous medical history. The ancient book on pulse diagnosis (the *Dapakani*) stresses that the physician taking the pulse needs to be clear in mind and body in order to do so. He or she will need:
1. a good moral attitude and diligent observation of precepts,
2. a clear spiritual and physical body,
3. expertise in aetiology and pathology,
4. firm concentration,
5. knowledge of the various qualities of pulse and their meaning.

Five-elements pulse diagnosis
A normal pulse is neither too big nor too small, too fast nor too slow, neither too strong nor too weak. It can be felt in both superficial and deep positions. It indicates a state of balance between the five elements.

Specific qualities of the pulse reflect each element. *Pathavi* governs the hardness or softness of the pulse. *Apo* governs the blood in the artery; how well it flows depends on the state of *vayaw* and its force depends on *tejo*. The clarity of any quality in the pulse depends on *akasa*.

The pulse can be taken in any of the following five locations, according to the *Dapakani* text:
1. *maniokka* (between acupuncture point Bl 1 and Bl 2),
2. *bawin waidika* (at the apex of the heart, or the radial pulse),
3. *khanda panchaccaya* (around the navel),
4. *pathamalawhita* (in the inguinal [groin] region),
5. *manizawtara* (below the medial malleolus).

1. *Maniokka* is located in the hollow at the medial end of the eyebrows, near the supraorbital notch. Take this pulse with the tip of the right thumb and index finger. If the pulse can be felt clearly, it indicates a normal state of *akasa*. If the

pulse cannot be felt or is slow, it indicates that *akasa* is weak and that *apo* is in excess.

2. *Bawin waidika* is, in theory, at the apex of the heart, but in practice the diagnostic location is on radial pulse. It can be felt by the tips of three fingers (index, middle and ring). Its qualities are classified according to the eight types of diseases. These are:

2(a) *ket-kha-la* pulse (H1) is rapid (*tejo*), strong (internal *pathavi* excess) and big (external *pathavi* excess);

2(b) 1st *vihtambhita* pulse (H2) is rapid, strong and thin (external *apo* and *pathavi*);

2(c) 2nd *vihtambhita* pulse (H3) is rapid, deep and irregular (weak and floating internal *pathavi*);

2(d) *prissava* pulse (H4) is rapid, deep, thin and weak (*ushna tejo*);

2(e) *singhahita* pulse (C1) is slow, big and weak (*sita tejo*);

2(f) 1st *byuhana* pulse (C2) is slow, big and thin (excess *sita pathavi* and *akasa*);

2(g) 2nd *byuhana* pulse (C3) is slow, deep, irregular and big (*sita vayaw* obstructed by *apo*);

2(h) *peggarana* pulse (C4) is slow, deep, thin, weak (excess of *akasa* and cold *vayaw*).

3. *Khandapanchacaya* is on or around the navel, and is felt by the right thumb. It indicates the state of *pathavi*. If this pulse can be felt clearly, then *pathavi* is normal. As the *pathavi* element is the residence of all other elements, if it is weak then all the others will be weak too.

4. *Pathamalawhita* can be felt in the inguinal region of the femoral artery. If it is clear and strong, it indicates that *tejo* is normal.

5. *Manizawtara* is below the medial malleolus (near acupuncture point Ki 6). It indicates the state of the relationship between *pathavi* and *apo*. If it is weak, *pathavi* and *apo* will be in excess. If the pulse is strong and rapid it indicates heat rising. If weak, it indicates that cold is descending.

Section 2.2: *Netkhattanaya* and *Weizzadharanaya*

Netkhattanaya

In this system of medicine, astrological calculations are used to plan diet, to diagnose certain diseases and to give a prognosis of the course of disease and the life expectancy of an individual patient. It is based on a reading of the placement of the planets at the time and place of the patient's birth compared to their current position. The theory is complex and sophisticated and there are no longer many practitioners.

Weizzadharanaya

This system of medicine has three major components:

1. Metallic medicine: is an application of ancient alchemy. It groups metals into two kinds:
(a) metals which represent the physical world:
(1) lead (black), (2) lead (brown), (3) iron, (4) tin, (5) mercury, (6) gold, (7) pure silver, (8) aluminium, (9) zinc;
(b) metals which represent the spiritual world:
(1) arsenic, (2) orpiment (yellow arsenic), (3) sulphur, (4) ammonium chloride, (5) copper sulphate, (6) cinnabar (mercuric sulphide), (7) boron, (8) alum, (9) potassium nitrate.

The basic method is to prepare drugs comprising a mixture of these metals, with some from each group. The mixture is put in an earthen container, sealed with mud and heated. Each formulation requires a different temperature and heating time and is then combined with liquids, herbal medicines or honey.

2. Psycho-spiritual therapy: consists of the recitation of Buddhist teachings in Pali verse; these will strengthen the patient by aiding his or her spiritual well-being and will ward off pathogens.

3. Power of the physician: there is an old saying that 'If the physician observes the precepts, then even the water he gives is medicinal'. In this system the physician has to undertake mental and spiritual purification through daily meditation and observation of the five precepts. Such practices give extraordinary power through intense concentration (called *samadhi*).

Section 3: Traditional Myanmar herbal medicine

Herbal medicine is the most reliable system of TM medicine. There are numerous forms of oral medicine, which are available in: (1) powder form, (2) pill or tablet form, (3) liquid (as a medicinal brew), (4) application or embrocation, (5) extracted juice.

All of the above preparations are still in use now. These days some pharmaceutical companies produce their medicines in capsule or caplet form, but most people do not like to take medicine in capsule form, as they believe that only the powder form and pills are an indication of genuine TM medicine.

1. Powder form

The powder form is the most useful, and is easier to take and to change into a pill or a tablet. The powder form is widely used in TM medicine. Making a powder is not difficult. In olden days, they were made by pounding dried herbs with mortar and pestle, but now grinding machines are used for the mass production of TM medicines. There are two main methods of formulation. The first one is to separately pound or grind each herb and then mix all the ingredients of the formulation respectively. The second method is to grind all ingredient herbs together.

There are numerous formulations in TM medicine. For township medical clinics forty-seven kinds of formulations are permitted to be prescribed, and eighty kinds of formulations are permitted to be prescribed in TMM hospitals. Other empirical remedies are also accepted for use in TMM hospitals by the prescription of physicians.

2. Making pills and tablets

The pounded substance must be mixed with sterilised drinking water to get it into the condition of a paste, which can then be formed into pills by hand (but a pilling machine can be used for mass production). For tableting, the powder must be mixed with starch (glue), then dried and made into granules and pressed by a tableting machine.

3. Medicinal brew

Some fresh herbal plants are boiled singly and taken for some diseases. This system has been used widely among our people and the knowledge has been passed on from one person to another for many years. Traditional Myanmar Medicine Practitioners use medicinal brews mixed with the oral medicine that they prescribe to get a dynamic efficacy. Some herbal plants are incredibly effective for some diseases, such as for diarrhoea and hepatitis types A, B, and C. We have many traditional single plant remedies.

4. Topical application/embrocation

There are numerous forms of topical applications. The most useful forms are (1) powder, and (2) ointment. Topical applications can be used for reducing swelling, inflammation, oedema, pain, stiffness of muscle joints, to promote union of fractured bones, and for healing open wounds.

Powder form

Some herbal plants and minerals, for example ginger (*Zingiber barbatum*) and camphor, are used for topical application, especially in musculo-skeletal problems. TMM practitioners believe that using topical application can unite a fractured bone quickly and reduce swelling more effectively than a plaster of Paris cast.

Ointment/liquid form

The liquid form can be formulated using mustard oil or sesame oil, camphor, beeswax and peppermint. It is used for muscle stiffness, musculo-skeletal problems, stiff joints and pain relief.

5. Extracted Juice

Extracted juices are also widely used in TMM. Some medicines can be mixed with the extracted juice of green medicinal plants to get more efficacy. For example, the extracted juice of the *Adhatoda vasica* leaf can be mixed with anti-bleeding medicines. Some practitioners use the extracted juice of medicinal plants in making pills to get more efficacy and to get a stickier consistency.

Section 4: Different treatment systems

In Traditional Myanmar Medicine there are also other treatment systems that may be used singularly or in combination with herbal medicine. These include (1) massage treatment, (2) hot fomentation (heat compress), (3) induced perspiration.

4.1. Massage treatment

There are many techniques of massage in the world, but Myanmar massage treatment is somewhat different. It is based on blood circulation and the five-elements theory. Myanmar also has massage for relaxation. Ancient kings and their families had a custom of receiving relaxation massage in the afternoon and before bed. They used to appoint many masseurs and masseuses in the palace. However, massage treatment is different from relaxation massage, as it utilizes anatomical theory and pathology. The anatomical theory and the pathology are based on the trajectory of the blood and the tonification of *tejo* and *vayaw* elements in the body.

Blood trajectory

Modern medical science explains that blood circulation is impelled from the heart and then spreads throughout the body. But in traditional Myanmar massage treatment it is believed that there are also wind (*vayaw*) trajectories which operate along the arteries and support all the activities of the body. TMM locates the main area where *vayaw* resides as being in the umbilicus. It is called '*combanali*': *comba* means tortoise and *nali* is a network of nerves. All the energies (comprising the *tejo* and *vayaw* elements) spread out in four trajectories, two of which are for the upper potion of the body, the other two being for the lower potion of the body (the legs).

The four trajectories pass throughout the body, operating beside all the arteries and supporting all the activities of body, including physiological functions. In TM massage anatomical theory it is believed that the artery from the heart meets together with four trajectories at the centre of body, in the umbilical region, which spread throughout the body. The main four trajectories have sub-energy stations (like trigger points) at the armpits, the base of the neck and the inguinal (groin) region. These areas can boost up the blood circulation and all the activities of the body.

A brief explanation of TM massage anatomy

In TM massage anatomy it is believed that the human body has an anatomical structure that accords with the five-elements theory. These correspondences are:
1. all phalanges correspond to the *akasa* element,
2. both wrist and ankle joints correspond to *vayaw* (wind element),

3. both elbow and knee joints correspond to the *tejo* element,
4. both shoulder and hip joints correspond to the *apo* (water) element,
5. both cervical and limbus sacral joints correspond to the *pathavi* (earth) element.

According to this concept, the massage treatment technique must be started at the *akasa* joints (phalanges), so that all the subsequent manoeuvres stimulate the *tejo* and *vayaw* elements and spread throughout the body. The final manoeuvre is massage on the 'main-station' area in the umbilical area. Practising this system leads to the regulation of the blood and *vayaw* (wind) circulation, according to the five-elements system. On the other hand, there are numerous massage points according with other positions, such as the seven massage points the on foot, three massage points on the leg, five knee points, five thigh points and seven lumbar points.

An example of treatment for cervical problems
Cervical problems can manifest in many signs and symptoms, such as neck pain, stiffness, pain in the scapular region, pain and tingling along the arm, and numbness of the fingers. This may be due to a localized blockage of the *vayaw* (wind) element through excess of the *pathavi* (earth) element, through deposition and cohesiveness at the cervical vertebral joints. The treatment must be started by massage from the phalanges (finger joints) to the shoulder, gradually, using the thumb pad. Then massage the base of the neck and the scapular regions, gradually going up to the nape of the neck. Finally, lift up the occiput with both hands and press both 'sub-station' areas of the neck for one minute.

4.2. Hot fomentation (hot compress)
In Myanmar hot fomentation is called *kyat htoat*. For this treatment some medicinal herbs are put together in a cotton sheet and tightly wrapped with a string. The herbs can be changed in accordance with different diseases to be treated. The ingredient herbs are different from those used in Ayurveda. Hot fomentation can eliminate bad blood, submerged wind, and wastage from physiological functions. The application of hot fomentation can promote the development of the *tejo* and *vayaw* elements, which cause warming of the blood and the promotion of blood and *vayaw* circulation. It can also disperse the cohesiveness of the *apo* element in the skin.

Indication
Hot fomentation can be used once a day for many problems, including pain reduction, arthritis, swelling joints, stiffness in joints, lower-back pain, sciatica, stiff muscles, body ache, numbness, tingling in the limbs, treating stroke, post-traumatic stiffness and deformities. This procedure can also be used at home as

a general household therapy.

Precautions
Patients with severe hypertension, tachycardia, severe diabetes, haemophilia and red eyes should not be treated with hot fomentation.

Varieties
There are three kinds of hot fomentation. These are:
1. oil- or butter-based hot fomentation,
2. fumigation-based hot fomentation,
3. direct heating-based hot fomentation.

The use of oil-based hot fomentation is very common in TM medical hospitals and clinics.

The ingredients of a package for hot fomentation
There are numerous kinds of herbs in a package for hot fomentation, but the most useful herbs are commonly used for musculo-skeletal problems. These are:
1. *Ricinus communis* (leaf) - 2gms
2. *Datura alba* (leaf) - 2gms
3. *Vitex trifolia* (leaf) - 2gms
4. *Croton oblongata* (leaf) - 2gms
5. *Citrus lemonum* (leaf) - 2gms
6. *Blumea balsamifera* (leaf) - 2gms

Total weight: 12gms

Preparation of a package for hot fomentation
1. Collect the leaves listed above and cut into pieces (each leaf must weigh 2 gms).
2. Put the leaves into a frying pan and roast with enough heat to cause withering.
3. Then put the leaves into a cotton sheet (1 sq. ft. in size) and wrap tightly with a string.
4. Put 4 inches of water in a steel pot (9 inches radius and 6 inches high) and cover with a steel tray (12 inches wide).
5. Put about 3 tablespoons of the hot fomentation oil on the tray and place the package on the tray.
6. Heat the pot to boil the water.

The package becomes hot when the water boils; then it is ready to use.

Making oil for hot fomentation
Sesame oil or mustard oil must be used for this medicine. 300 cl of oil is boiled

and then the extracted juice of collected leaves (about 12 gms) is added to the boiling oil (the oil becomes greenish-yellow colour) to make one package. Care must be taken so that the foam does not overflow. The mixture is then left to cool down.

The application of hot fomentation

1. Rubbing
This is the first system of application. Before the use of the preparation, the TM practitioner should check the temperature of the preparation with his or her hand: the temperature should be about 50°C to 60°C. The physician should then warm his or her hand by rubbing the package, thus reducing the temperature difference between his or her hand and the skin of the patient. Then rub the treatment area of the patient, first with the hand, and then, pressing lightly, with the package containing the preparation.

2. The press and lift method
This procedure entails quickly pressing and lifting the application in various places in the treatment area.

3. Rubbing with pressing
This procedure is to begin rubbing, then pressing lightly seven times, and then pressing with force.

4. Pressing with force
This procedure is used to treat problems of bones and joints.

5. Circled pressing
This procedure is used for localised problems. It involves pressing and lifting clockwise.

6. Spiral pressing
This procedure is used on large muscle areas for problems in the upper back, mid-back and lower back.

Sometime, all procedures can be used at the same time, step by step. The duration of a hot compress should be for ten to fifteen minutes for the best results.

4.3. Inducement of perspiration

Inducement of perspiration is one of treatment methods of TMM. It is used in combination with herbal medicine (oral medicine) for the treatment of disease, as well as for protection from disease, and can be used for general family healthcare. This treatment can be used for fever, oedema, body ache, obesity, heaviness of limbs and some muscle injuries. The herbs used as ingredients play an important role in the perspiration treatment. These herbs are different according to which disease is being treated, but most TMM practitioners use only a few formulations to induce perspiration. Some herbs help to promote perspiration. These are:

1. bamboo leaf - 3gms
2. *Vitex trifolia* - 3gms
3. citrus lemons - 3gms
4. *Croton oblongata* - 3gms
Total weight: 12gms

The procedure of inducement of perspiration

First of all, leaves are put into a pot (about 12 inches wide and 9 inches high) and mixed with water, filling $^2/_3$ of the pot. Then heat the water until it boils. The patient should sit on a chair and put the pot between his or her legs, with a blanket covering the whole body. Then remove the cover of pot. The duration of the procedure should be about twenty minutes per session. The main purpose is to induce sweating all over the body. If sweat comes from the forehead and the nose, then the induced perspiration has been successful. The pot should then be moved away and the patient allowed to sit (covered with a blanket) on the chair for five minutes. Then rub the sweat with a towel and change the cloth. After that, remove the blanket and ask the patient to drink warm water, which helps sweating. Inducing perspiration can be done not only for the entire body but also for part of the body. It is not difficult to do, just cover the part of the body where treatment is needed. There are now, also, modern apparatuses to induce perspiration.

Section 5: Integration of TMM with TCM and acupuncture

There is now significant evidence of very good results, in the treatment of some diseases, from the integration of TMM treatment with traditional Chinese medicine (TCM) and acupuncture, especially for facial palsy, stroke, cardiac problems, post-traumatic stiffness, sciatica, tumour in brain, and some infectious diseases.

TMM has comprehensive systems of treatment, through herbal medicine and other methods. But some diseases are stubborn. However, the use of alternative treatments can in some cases produce rapid physiological changes and quickly reduce problems. For example, some diseases which have been treated by allopathic medicine for a long time but which show little sign of improvement, may be quickly diminished by using TMM. Similarly, if a long period of treatment by TMM cannot produce an improvement, then the use of acupuncture may cause rapid improvement.

Research was conducted between 2008 and 2010 on a total of fifty-seven patients who were treated in my clinic. There were twelve patients with facial palsy, twenty-one patients with chronic lower-back pain and sciatica, fourteen patients who had had a stroke, three patients with a brain tumour, four patients with epilepsy, and three patients with cardiac problems. The results indicate that integrated treatment can reduce the duration of treatment, and rapid improvement can be achieved. Some other problems, such as post-traumatic stiffness and arthritis, have also been treated with an integrated approach, but as the positive results of integrated treatment have already been clearly demonstrated, there is no need of research to demonstrate the efficacy of treatment in these cases.

Most cases of integration treatment are with acupuncture and TMM. The procedure of treatment can be as follows: firstly, application of needles; secondly, application of TMM massage (if needed); thirdly, localised application with TMM (if needed); finally, TMM herbal medicine (oral medicine).

Traditional Myanmar medical applications are particularly effective for swelling, oedema, and inflammation of joints. The most useful formulations of application are as follows:

TMF 29 *Thet ba ze ya se*
1   *jin chaut* (dried ginger) 1092 gms
2   *pitchin thee* (long pepper, *Piper longum*) 276 gms (fruit)
3   *nga yot* (black pepper, *Piper nigrum*) 260 gms
4   *pa yoke* (camphor) 68 gms
5   *karaway* (*Cinnamomum tamala*) 4 gms (leaf)
6   *karaway thee* (*Cinnamomum tamala*) 4 gms (fruit)
7   *samonsa pa* (*Pimpinella ainsum*) 4 gms (grain)

8   samon phwe (*Peucedanum*) 4 gms (stalk)
9   samon phyu (*Thymus vulgaris*) 4 gms
10 samon ni (*Lepidium sativum*) 4 gms (seed)
11 samon nyo (*Apium graveolens*) 4 gms (grain)
12 let char (*Borex*) 4 gms
13 zar pwint (*Myrista fragrans*) 4 gms
14 thitgabo (*Cinnamomum*) 4 gms (bark)
15 lay hnyin (clove) 4 gms
16 phalar (*Elettaria cardamomum*) 4 gms
17 pitchin myit (*Piper longum*) 4 gms (root)
18 kya su (*Terminalalia citrina*) 4 gms (fruit)

This formulation (above) can produce 'heat' in the area where it is applied. It can be used not only as an application but also as an oral medicine. It can reduce pain, swelling and inflammation. Although this formulation produces heat, it can also eliminate heat.

The formulation below, TMF 43, can reduce heat in the area where it is applied and can also reduce pain, swelling, inflammation, itching skin, and a rash. Every TMM practitioner use these two formulations, mixed in equal proportion, for swelling, oedema and traumatic injuries.

TMF 43
1   gway dauk myit (*Dragea volubilis*) 16 gms
2   na nwin nwe 16 gms
3   myet hmway 16 gms
4   pan yin (*Vetivera zizanoides*) 16 gms
5   pan le (*Woodfordia fruticosa*)16 gms
6   pannu (*Saussurea spicata*) 16 gms
7   pan u (*Kaempferia*) 16 gms
8   kant palu u (*Valeriana wallichi*) 16 gms
9   pe nant thar (*Trigomella foenum graecum*) 16 gms
10 u pa tha ka (*Hemidesmus indicus*) 16 gms
11 thit gya bo (*Cinnamomum cassia*) 16 gms
12 gant gaw wut san (*Mesua ferrea*) 16 gms
13 nant thar phyu (*Santalum album*) 16 gms
14 nant thar ni (*Pterocarpus santalinus*) 16 gms
15 dantaku ni (*Pterocarpus santalinus*) 16 gms

Case history 1
A woman aged fifty-five was suffering from anxiety, palpitation, insomnia, sudden feelings of weakness, had a bad smell her nose, and had frequent mild fits. During the previous five years she had suffered from a brain tumour and

had been operated on by a surgeon. She had been asked to return to be re-examined every six months but had not complied.
Clinical manifestations:
Blood pressure: 162/92 mm Hg.
Tongue: thick yellow coating, purple, red tip.
Pulse: wiry

Acupuncture treatment principle:
4 gates and Sp 9, 6 were used for dispersing dampness and eliminating heat. SI 3 and Bl 62 were used to subdue interior wind, and to nourish the marrow and the brain.
TM medicine:
The pathology: *tejo* and *vayaw* elements are surrounded and blocked by the *apo* element, so that the trapped *tejo* and *vayaw* elements are attempting to get out. This condition results in the expression of the symptoms, like palpitation, insomnia, and convulsion.
Treatment principle: to subdue interior heat (the *tejo* element) and regulate the *vayaw* element. Herbal medicine was used to cool, through sweat and the mildly bitter taste of the medicine. TMM massage was also used. After the first treatment, sleep became
good and the frequent fits became less frequent, from seven times a day to three times a day.

The second visit: 4 gates, Gb 39 and SI 3; blood pressure:150/90 mm Hg.
The third treatment: Sp 9/6 and Lu 7, K 6, H 7 and neurogates (ear acupuncture points). After the third treatment the fits stopped. TM herbal medicine was continued, and acupuncture was used when the fits reappeared. This treatment could have enhanced her life. Unfortunately, she passed away in August 2010 due to cerebral haemorrhage.

Case history 2
A man aged eighty-two presented with signs and symptoms of paraparesis (mild, left-sided hemiplegia) due to a small intracranial osteoma, arising from the inner table of the (left) frontal bone. He had been treated with allopathic medicine for one month, but had not improved.
Clinical manifestation:
Blood pressure: 120/85 mmHg
Tongue: red, dry, with a thin yellow coat
Pulse: superficial, thin

Acupuncture treatment principle:
First treatment: SI 3 and Bl 62 were used to nourish the brain and the marrow.

Sp 6, and K3 were used for K-yin tonification, 4 gates for regulating meridians. Second treatment: SI 3 Bl 62 GB 39 and Sp 6, after second treatment, the patient could walk without support.

TM medicine:
Topical application with ginger and *linnay* (sweet flag) powder with warm water once a day and massage. This problem started from a weak *vayaw* element because of lack of the supportive earth (*pathavi*) element.

Treatment principle: tonification for the earth (*pathavi*) and wind (*vayaw*) elements. The taste of medicine should be bitter and somewhat sweet. The patient needs to take TM drugs for long period of time to reduce the intracranial osteoma.

Case history 3
A woman aged sixty-seven, presented with chronic signs and symptoms of vertigo, anxiety, heaviness, and hypertension. She had been treated by an allopathic doctor for seven days, but had not improved. She was in an anxious state due to her son working away from home.

Clinical manifestation:
Blood pressure: 190/85 Hg
Tongue: thick yellow coating, red tip (Lu, Ht area)
Pulse: full, tight

Treatment principle: Sp 4, p 6 and Lu 7 KG were used to calm her mind and permit good sleep, and to reduce systolic pressure. Sp 9 was used for dispersing dampness.

TMM principle:
She had an excess of internal and external *apo* and *pathavi* elements and was in a hot state. The treatment principle was to disperse the *apo* and *pathavi* elements and reduce internal heat. The taste of the medicine was to be sweet and sour. After one visit her blood pressure went down to 150/80Hg and her vertigo reduced.

Section 6: The Department of Traditional Myanmar Medicine

Traditional Myanmar Medicine (TMM) was officially set up as a branch of the health service in 1972. It originated as a division under the Department of Health, managed by an Assistant Director who was responsible for the development of the service under the technical guidance of the State Traditional Medicine Council and became the focal point for all activities related to TM. This division of TM was able to establish the Institute of Traditional Medicine in Mandalay, along with a twenty-five-bed teaching hospital and a sixteen-bed pilot research hospital in Yangon. Patient services were offered to the public by TM clinics, which were gradually increased to 110 by 1988. Due to its popularity and good acceptance by the community, the government finally decided to upgrade the division to become a separate Department of Traditional Myanmar Medicine in August 1989. The new organizational set-up includes one Director General, two Deputy Director Generals, four Directors, Deputy Directors and Assistant Directors. There are ten divisions in the Department to manage the activities of developmental objectives, mainly in the areas of training, research, drug production, management, finance and treatment care. At the moment, the total strength of Department is set to increase yearly.

Mission of the Department
1. To develop TMM Medical service down to grass-roots level in the community, narrowing the gap between urban and rural population.
2. To function as a complementary arm of the health care system of the country in close coordination with the medicine programme and Primary Healthcare Programme of the Department of Health.
3. To develop traditional medicinal resources needed for TMM.
4. To conduct systematic research into TM drugs, service and practice.
5. To develop TMM to become acceptable to the international community.

The regulation of practitioners of Myanmar medicine
If a person wants to be a practitioner in Myanmar, there must be completion of the following qualifications from Department of Traditional Medicine, Ministry of Health. These are:
(a) graduation from the Traditional Medical Institute or Traditional Medicine University;
(b) completion of one year of training for becoming a Traditional Medicine Practitioner by the Department of Traditional Medicine, Ministry of Health;
(c) completion of the qualifying examination held by the traditional Myanmar Medical Practitioners Board of the Department of Traditional Medicine, Ministry of Health.
   At the present time, there is only one way to be a practitioner, and that is

to be a graduate from the University of Traditional Medicine. If a person has completed all of the above qualifications, he or she may apply to the council for registration as a traditional medical practitioner. A person who has graduated from the Traditional Medical Institute or University may submit for government service at the Department of Traditional Medicine, Ministry of Health. A person who possesses only two certificates cannot submit for government service, but he or she can work in private clinics.

Duties and registration of a Traditional Medical Practitioner
A traditional medicinal practitioner shall:
(a) abide by the rules, procedures, order and directives issued by the Ministry of Health;
(b) abide by and observe the rules of conduct and discipline prescribed by the council;
(c) pay the annual fees prescribed periodically by the council;
The traditional medicine practitioner has the right to:
(a) be assigned the duty of, be elected as, and to elect, a council member;
(b) practise traditional medicine as a profession;
(c) tender advice to the council;
(d) submit his/her grievances to the council.

Functions and duties of the Ministry of Health
The Ministry of Health:
(a) may appoint and assign duty to any officer from the Department of Traditional Medicine to act as a registration officer for the purpose of TMM council law;
(b) shall prescribe the duties and powers of a registration officer.
The Ministry of Health may, in respect of issuing certificates for practice and for taking action against medical practitioners who practise by applying traditional medicine of any foreign country, assign duties and responsibilities to the Department of Traditional Medicine.

The formation of the Traditional Medicine Council
The Ministry of Health, by the approval of the government has formed the Traditional Medicine Council comprising the following persons:
(a) Director General (Chairman);
(b) four Traditional Medical Practitioners assigned for duty as members, by the Ministry of Health;
(c) five persons elected by a member of the Traditional Medical Practitioners, from among themselves;
(d) an officer assigned by the secretary for the Ministry of Health.

The role of Traditional Medicine practitioners

Myanmar Traditional Medicine practitioners have been providing medical care for people since time immemorial. Most practitioners use the *Desa* system and *Beithizza* (Ayurvedic) system, while a only few use the *Weizzadhara* system and the *Netkhatta* (medical astrology) system.

Each traditional medicine practitioner provides patients with skilled knowledge of each system and provides the entire nation with comprehensive traditional medical services through the existing healthcare system of Myanmar, either via government- owned traditional medicine hospitals and clinics or through their private clinics. On the other hand, all the traditional medical practitioners are organized as one organization by the department of traditional medicine, which is called the 'Myanmar Traditional Medicine Practitioners Association' (M.T.M.P.A.).

The Myanmar Traditional Medicine Practitioners Association (M.T.M.P.A.)

The M.T.M.P.A. was established in June 2003 and is one of the NGOs which operates in Myanmar. Most TMPs participate in all the activities of the M.T.M.P.A. The M.T.M.P.A. supports some activities of the Department of Traditional Medicine, especially in the public healthcare sectors, and also in other activities related to TM science. The M.T.M.P.A. lays down aims and objectives and implements them in yearly programs.

The aims and objectives of the M.T.M.P.A.

The organization's aims and objectives are:
1. to provide medical services by the collective effort of Traditional Medicine Practitioners;
2. to modernize traditional Myanmar Medicines in conformity with scientific methods;
3. to revive rare and extinct ancient formulations and therapies of Traditional Myanmar Medicine;
4. to promote continuous medical education for TMM practitioners;
5. to cooperate with the relevant government departments, organizations and international organizations of traditional medicine.

Activities undertaken by M.T.M.P.A. (Mandalay region) from 2003 to 2010

The activities undertaken included:
1. operating community outreach programmes for remote areas;
2. providing continuous medical education with the cooperation of University of Traditional Medicine (research papers are read every month, medical education for is provided for rural people);
3. reading research papers at TM Medical Practitioners conferences yearly;

4. running refresher training courses for TM medical practitioners;
5. participation in relief work and providing medical services to victims in areas of natural disaster.

Organizations involved in TMM
Organizations working with TMM include the Ministry of Health, the Department of Traditional Medicine, the Traditional Medicine Council, private and state-run hospitals and clinics, and the Myanmar Traditional Medicine Practitioners Association (M.T.M.P.A.), of which there are staff and branch associations at state, regional, and district offices. There are also channels of cooperation between the M.T.M.P.A. and the Ministry of Health, and between practitioners of TMM and traditional Chinese medicine.

References
Country Report on Traditional Myanmar Medicine (2nd workshop, Shanghai) (2010). Naypyidaw: Department of Traditional Medicine, Ministry of Health.
A Handbook of Traditional Medicine (2008). Naypyidaw: Department of Traditional Medicine, Ministry of Health.
The History of Traditional Medicine in Myanmar (1975). Naypyidaw: Department of Traditional Medicine, Ministry of Health.
Hudson, Bob (2004). *The Origin of Bagan: The Archaeological Landscape of Upper Burma to AD 1300* (unpublished PhD thesis). University of Sydney. http://acl.arts.usyd.edu.au/~hudson/bobhpage.htm
San Shwe (2006). 'The Culture of Vishnu Old City (Beikthano)', *Uncovering Southeast Asia's Past*. Singapore: NUS Publishing, pp. 272–83.
Sinphyukyune Aung Thein (1986). *The Traditional Custom of Myanmar Tattoo*. Yangon: Sar Pay Beik Mam.
Than Tun (2003). *Myanmar Terracottas (Pottery in Myanmar & Votive Tablets of Myanmar)*. Yangon: Chit Sayar Sarpay.

www.ingramcontent.com/pod-product-compliance
Lightning Source LLC
Chambersburg PA
CBHW070434180526
45158CB00017B/1220